The Institution of Civil Engineers

The New Engineering Contract
Option B

A form of contract for a priced
contract with bill of quantities

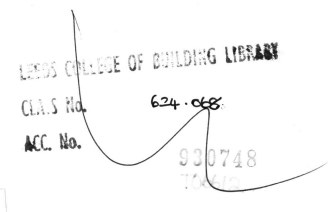

 Thomas Telford, London

Published for the Institution of Civil Engineers by Thomas Telford Services Ltd, Thomas Telford House, 1 Heron Quay, London E14 4JD

The New Engineering Contract is published as a series of documents of which this is one.

ISBN (series) 0 7277 1664 6

ISBN (this document) 0 7277 1945 9

Consultative edition 1991
First edition 1993

British Library Cataloguing in Publication Data for this publication is available from the British Library.

Printed and bound in Great Britain by Staples Printers Rochester Ltd, Rochester, Kent.

CONTENTS

In this contract the core clauses are the NEC core clauses and the clauses set out in the NEC as main option clauses: Option B. The latter are included in sequence and are printed in **bold type** in this contract.

ACKNOWLEDGEMENTS

The New Engineering Contract has been produced by the Institution of Civil Engineers through its New Engineering Contract Working Group.

The New Engineering Contract has been designed and drafted by Dr Martin Barnes of Coopers and Lybrand with the assistance of Professor J. G. Perry of The University of Birmingham, T. W. Weddell of Travers Morgan Management, T. H. Nicholson, Consultant to the Institution of Civil Engineers, A. Norman of the University of Manchester Institute of Science and Technology and P. A. Baird, Corporate Contracts Consultant, Eskom, South Africa.

The members of the New Engineering Contract Working Group are

R. L. Wilson, CBE, BSc(Eng), FEng, FICE (Chairman)
M. W. Abrahamson, BA, LLB, FCIArb
P. A. Baird, BSc, CEng, FICE, M(SA)ICE, MAPM
M. Barnes, BSc(Eng), PhD, FEng, FICE, FCIOB, CBIM, FAPM, FInstCES, ACIArb
J. A. Chandler, MA, CEng, FICE, FCIArb
L. T. Eames, BSc, FRICS, MCIOB
F. Griffiths, CEng, FIEE, FInstPS, FBIM
J. Halliday, CEng, MICE
K. Lumb, FRICS, ACIArb
W. S. McAlonan, MSc, FEng, FICE, FIHT
T. H. Nicholson, BSc, CEng, FICE (Secretary)
A. Norman, BSc, MSc, CEng, MICE, MAPM
Professor J. G. Perry, MEng, PhD, CEng, MICE, MAPM
T. W. Weddell, BSc, CEng, DIC, FICE, FIStructE, ACIArb

I. M. H. Moore, CBE (Director External Affairs, ICE)
J. J. Lewis (Project Manager, June 1991 – January 1992)
R. F. Bell, BSc, CEng, FICE (Project Manager from January 1992)

The Institution of Civil Engineers also acknowledges the considerable contributions made to the New Engineering Contract by

N. G. Bunni, BSc, MSc, PhD, CEng, FIEI, FICE, FCIArb
S. C. McCarthy, BE, MSc, PhD, MIEI

SCHEDULE OF OPTIONS

The following secondary options should be considered. It is not necessary to use any of them. Any combination may be used.

Option G	Performance bond
Option H	Parent company guarantee
Option J	Advanced payment to the *Contractor*
Option K	Multiple currencies
Option L	Sectional Completion
Option M	Limitation of the *Contractor*'s liability for design to reasonable skill and care
Option N	Price adjustment for inflation
Option P	Retention
Option Q	Bonus for early Completion
Option R	Delay damages
Option S	Low performance damages
Option T	Changes in the law
Option U	Special conditions

THE NEW ENGINEERING CONTRACT

CORE CLAUSES

1 General

Actions	**10**	
	10.1	The *Employer,* the *Contractor,* the *Project Manager,* the *Supervisor* and the *Adjudicator* shall act as stated in this contract.
Identified and defined terms	**11**	
	11.1	In these conditions of contract, terms identified in the Contract Data are in italics and defined terms have capital initials.

11.2 (1) The Parties are the *Employer* and the *Contractor.*

(2) Others are people or organisations who are not the *Employer,* the *Project Manager,* the *Supervisor,* the *Adjudicator,* the *Contractor,* or any employee, Subcontractor or supplier of the *Contractor.*

(3) The Contract Date is the date when this contract came into existence.

(4) To Provide the Works means to do the work necessary to complete the *works* in accordance with this contract and all incidental work, services and actions which this contract requires.

(5) Works Information is information which

- specifies and describes the *works* or
- states how the *Contractor* Provides the Works and is either

  - in the documents which the Contract Data states it is or
  - in an instruction given in accordance with this contract.

(6) Site Information is information which

- describes the Site and its surroundings and
- is in the documents which the Contract Data states it is.

(7) The Site is the area within the *boundaries of the site* and the volumes above and below it which are affected by work included in this contract.

(8) The Working Areas are the *working areas* unless later changed in accordance with this contract.

(9) A Subcontractor is a person or corporate body who has a contract with the *Contractor* to provide part of the *works* or to supply Plant and Materials which he has wholly or partly designed specifically for the *works.*

(10) Plant and Materials are items to be included in the *works.*

(11) Equipment is items provided by the *Contractor* and used by him to Provide the Works and which the Works Information does not require him to include in the *works.*

(12) The Completion Date is the *completion date* unless later changed in accordance with this contract.

(13) Completion is the date, decided by the *Project Manager*, when the *Contractor* has done all the work which the Works Information states he is to do by the Completion Date and has corrected notified Defects which would have prevented the *Employer* using the *works*.

(14) The Accepted Programme is the latest programme accepted by the *Project Manager*. If no programme has been accepted by the *Project Manager*, a programme identified in the Contract Data is the Accepted Programme.

(15) A Defect is

- a part of the *works* which is not in accordance with the Works Information or
- a part of the *works* designed by the *Contractor* which is not in accordance with

  - the applicable law or
  - the *Contractor*'s design which has been accepted by the *Project Manager*.

(16) The Defects Certificate is either a list of Defects notified before the *defects date* which the *Contractor* has not corrected or, if there are no such Defects, a statement that there are none.

(17) Risk Transfer is the earlier of the date when the *Project Manager* certifies that the *Employer* has taken over the *works* and the date when he certifies termination.

(18) The Fee is the amount calculated by applying the *fee percentage* to the amount of Actual Cost.

(21) The Prices are the lump sums and the amounts obtained by multiplying the rates by the quantities for the items in the *bill of quantities* unless later changed in accordance with this contract.

(25) The Price for Work Done to Date is the total of

- **the quantity of the work which the *Contractor* has completed for each item in the *bill of quantities* multiplied by the rate and**
- **a proportion of each lump sum which is the proportion of the work covered by the item which the *Contractor* has completed.**

(28) Actual Cost is the cost of the components in the Schedule of Cost Components whether work is subcontracted or not.

Interpretation 12

12.1   In this contract, except where the context shows otherwise, words in the singular also mean the plural and the other way round and words in the masculine also mean the feminine and neuter.

Communications 13

13.1   Each instruction, certificate, submission, proposal, record, acceptance and notification which this contract requires is communicated in a form which can be read, copied and recorded. Writing is in the *language of this contract*.

13.2   A communication has effect when it is received at the last address notified by the recipient for receiving communications or, if none is notified, at the address of the recipient stated in the Contract Data.

13.3   A reply to a communication between the *Project Manager* or the *Supervisor* and the *Contractor* to which this contract requires a reply is made within the *period for reply*.

13.4   The *Project Manager* replies to a communication submitted or resubmitted to him by the *Contractor* for acceptance. If his reply is not acceptance, he states his reasons and the *Contractor* resubmits the communication within the *period for reply* taking account of these reasons. A reason for withholding acceptance is that more information is needed in order to assess the *Contractor*'s submission fully.

13.5 The *Project Manager* may extend the *period for reply* to a communication if the *Project Manager* and the *Contractor* agree to the extension before the reply is due. The *Project Manager* notifies the extension which has been agreed to the *Contractor*.

13.6 The *Project Manager* issues his certificates to the *Employer* and the *Contractor*. The *Supervisor* issues his certificates to the *Project Manager* and the *Contractor*.

13.7 Information which this contract requires to be notified is communicated separately from other communications.

The *Project Manager* and the *Supervisor* **14**

14.1 The *Project Manager*'s or the *Supervisor*'s acceptance of a communication from the *Contractor* or of his work does not change the *Contractor*'s responsibility to Provide the Works or his liability for his design.

14.2 The *Project Manager* and the *Supervisor*, after notifying the *Contractor*, may delegate any of their actions and may cancel any delegation. A reference to an action of the *Project Manager* or the *Supervisor* in this contract includes an action by his delegate.

14.3 The *Project Manager* may give an instruction to the *Contractor* which changes the Works Information.

14.4 The *Employer* may replace the *Project Manager* or the *Supervisor* after he has notified the *Contractor* of the name of the replacement.

Adding to the *working areas* **15**

15.1 The *Contractor* may submit a proposal for adding to the Working Areas to the *Project Manager* for acceptance. Reasons for not accepting are that the proposed addition is not necessary for Providing the Works and that the proposed area will be used for work not in this contract.

Early warning **16**

16.1 The *Contractor* and the *Project Manager* give an early warning by notifying the other as soon as either becomes aware of any matter which could increase the total of the Prices, delay Completion or impair the performance of the *works* in use.

16.2 Either the *Project Manager* or the *Contractor* may instruct the other to attend an early warning meeting. Each may instruct other people to attend if the other agrees.

16.3 At an early warning meeting those who attend co-operate in

- making and considering proposals for how the effect of each matter which has been notified as an early warning can be avoided or reduced and
- deciding upon actions which they will take and who, in accordance with this contract, will take them.

16.4 The *Project Manager* records the proposals considered and decisions taken at an early warning meeting and gives a copy of his record to the *Contractor*.

Ambiguities and inconsistencies **17**

17.1 The *Project Manager* or the *Contractor* notifies the other as soon as either becomes aware of an ambiguity or inconsistency in or between the documents which are part of this contract. The *Project Manager* gives an instruction resolving the ambiguity or inconsistency.

Health and safety   **18**

18.1   The *Contractor* acts in accordance with the health and safety requirements stated in the Works Information.

Illegal and impossible   **19**
requirements   19.1   The *Contractor* notifies the *Project Manager* as soon as he becomes aware that the Works Information requires him to do anything which is illegal or impossible. If the *Project Manager* agrees, he gives an instruction to change the Works Information appropriately.

2 The *Contractor*'s main responsibilities

Providing the Works   **20**

20.1   The *Contractor* Provides the Works in accordance with the Works Information.

The *Contractor*'s design   **21**

21.1   The *Contractor* designs the parts of the *works* which the Works Information states he is to design. His design complies with the Works Information.

21.2   The *Contractor* submits the particulars of his design as the Works Information requires to the *Project Manager* for acceptance. Reasons for not accepting are that the design does not comply with the Works Information and that it does not comply with the applicable law. The *Contractor* does not proceed with the relevant work until the *Project Manager* has accepted his design.

21.3   The *Contractor* may submit his design for acceptance in parts if the design of each part can be assessed fully.

21.4   The *Contractor* indemnifies the *Employer* against claims, compensation and costs due to the *Contractor* infringing a patent or copyright.

21.5   The *Contractor*'s liability to the *Employer* for his design is limited after Completion of the whole of the *works* to the amount stated in the Contract Data.

Using the *Contractor*'s design   **22**

22.1   The *Employer* may use and copy the *Contractor*'s design for installing, constructing, re-installing or reconstructing the *works* unless otherwise stated in the Works Information and for other purposes as stated in the Works Information.

Design of Equipment   **23**

23.1   The *Contractor* submits particulars of the design of an item of Equipment to the *Project Manager* for acceptance if the *Project Manager* instructs him to. Reasons for not accepting are that the design of the item will not allow the *Contractor* to Provide the Works in accordance with

- the Works Information,
- the *Contractor*'s design which the *Project Manager* has accepted or
- the applicable law.

People   **24**

24.1   The *Contractor* either employs each key person named to do the job stated in the Contract Data or employs a replacement person who has been accepted by the *Project Manager*. The *Contractor* submits the name, relevant qualifications and experience of a proposed replacement person to the *Project Manager* for acceptance. A reason for not accepting the person is that his relevant qualifications and experience are not as good as those of the person who is to be replaced.

24.2   The *Project Manager* may, having stated his reasons, instruct the *Contractor* to remove an employee. The *Contractor* then arranges that, after one day, the employee has no further connection with the work included in this contract.

Co-operation **25**

25.1   The *Contractor* co-operates with Others in obtaining and providing information which they need in connection with the *works*. He shares the Working Areas with Others as stated in the Works Information.

Assignment and **26**
subcontracting 26.1   Neither Party assigns any benefit in a part or the whole of this contract. If the *Contractor* subcontracts work, he is responsible for performing this contract as if he had not subcontracted. This contract applies as if a Subcontractor's employees and equipment were the *Contractor*'s.

26.2   The *Contractor* submits the name of each proposed Subcontractor to the *Project Manager* for acceptance. A reason for not accepting the Subcontractor is that his appointment will not allow the *Contractor* to Provide the Works in accordance with this contract. The *Contractor* does not appoint a proposed Subcontractor until the *Project Manager* has accepted him.

Approval from Others **27**

27.1   The *Contractor* obtains approval of his design from Others where necessary.

Access to the work **28**

28.1   The *Contractor* provides access for the *Project Manager*, the *Supervisor* and others notified by the *Project Manager* to work being done for this contract and to stored Plant and Materials.

Instructions **29**

29.1   The *Contractor* obeys an instruction which the *Project Manager* or the *Supervisor* gives him and which is in accordance with this contract.

3 Time

Starting and Completion **30**

30.1   The *Contractor* does not start work on the Site until the first *possession date* and does the work so that Completion is on or before the Completion Date.

30.2   The *Project Manager* certifies Completion within one week of Completion.

The Accepted **31**
Programme

31.1   If the Accepted Programme is not identified in the Contract Data, the *Contractor* prepares and submits a programme to the *Project Manager* for acceptance within the period stated in the Contract Data.

31.2   Reasons for not accepting a programme (including a revised programme) are that the *Contractor*'s plans which it shows are not practicable and that the programme does not

- include the information which this contract requires
- represent the *Contractor*'s plans realistically or
- show realistic provisions for

  - float and other risk allowances
  - health and safety requirements
  - other requirements of the Works Information or
  - the procedures set out in this contract.

31.3   The Accepted Programme includes

- the *starting date*, *possession dates* and Completion Date
- for each operation, a method statement which identifies the Equipment and other resources which the *Contractor* plans to use
- planned Completion
- the order and timing of

  - the operations which the *Contractor* plans to do in order to Provide the Works and
  - the work of the *Employer* and Others either as stated in the Works Information or as later agreed with them by the *Contractor*

- the dates when the *Contractor* plans to complete work needed to allow the *Employer* and Others to do their work
- the dates when, in order to Provide the Works in accordance with his programme, the *Contractor* will need

  - possession of a part of the Site if later than its *possession date*
  - acceptances and
  - Plant and Materials and other things to be provided by the *Employer* and

- other information which the Works Information requires the *Contractor* to show on the Accepted Programme.

Revising the programme **32**

32.1   Each revised programme shows

- the actual progress achieved on each operation since the last revision and its effect upon the timing of the remaining work
- the effects of implemented compensation events and of notified early warning matters

- how the *Contractor* plans to deal with any delays and to correct notified Defects and
- any other changes which the *Contractor* proposes to make to the Accepted Programme.

32.2 The *Contractor* submits a revised programme to the *Project Manager* for acceptance

- within the *period for reply* after the *Project Manager* has instructed him to
- when the *Contractor* chooses to and, in any case,
- at no longer interval than the interval stated in the Contract Data from the *starting date* until Completion of the whole of the *works*.

Possession of the Site **33**

33.1 The *Employer* gives possession of each part of the Site to the *Contractor* on or before the later of its *possession date* and the date for possession shown on the Accepted Programme.

33.2 While the *Contractor* has possession of a part of the Site, the *Employer* gives the *Contractor* access to and use of it and the *Employer* and the *Contractor* provide facilities and services as stated in the Works Information. Any cost incurred by the *Employer* as a result of the *Contractor* not providing the facilities and services stated is assessed by the *Project Manager* and paid by the *Contractor*.

Instructions to stop or **34**
not to start work 34.1 The *Project Manager* may instruct the *Contractor* to stop or not to start any work and may later instruct him that he may re-start or start it.

Taking over **35**

35.1 Possession of each part of the Site returns to the *Employer* when he takes over the part of the *works* which occupies it. Possession of the whole Site returns to the *Employer* when the *Project Manager* certifies termination.

35.2 The *Employer* need not take over the *works* before the Completion Date if it is stated in the Contract Data that he is not willing to do so. Otherwise the *Employer* takes over the *works* not more than two weeks after Completion.

35.3 The *Employer* may use any part of the *works* before Completion has been certified. If he does so, he takes over the part of the *works* when he begins to use it.

35.4 The *Project Manager* certifies the date upon which the *Employer* takes over any part of the *works* within one week of the date.

Acceleration **36**

36.1 The *Project Manager* may instruct the *Contractor* to submit a quotation for an acceleration to achieve Completion before the Completion Date. A quotation for an acceleration comprises proposed changes to the Prices and the Completion Date and a revised programme.

36.2 The *Contractor* submits a quotation or gives his reasons for not doing so within the *period for reply*.

36.3 When the *Project Manager* accepts a quotation for an acceleration, he changes the Completion Date and the Prices accordingly and accepts the revised programme.

4 Testing and Defects

Tests and inspections **40**

40.1 This clause only governs tests and inspections required by the Works Information and the applicable law.

40.2 The *Contractor* and the *Employer* provide materials, facilities and samples for tests and inspections as stated in the Works Information.

40.3 The *Contractor* and the *Supervisor* notify the other of each of their tests and inspections before it starts and afterwards notify its results. The *Contractor* notifies the *Supervisor* in time for a test or inspection to be arranged and done before doing work which would obstruct the test or inspection. The *Supervisor* may watch any test done by the *Contractor*.

40.4 If a test or inspection shows that any work has a Defect, the *Contractor* corrects the Defect and the test or inspection is repeated.

40.5 The *Supervisor* does his tests and inspections without causing unnecessary delay to the work or to a payment which is conditional upon a test or inspection being successful. A payment which is conditional upon a *Supervisor*'s test or inspection being successful becomes due at the later of the *defects date* and the end of the last *defect correction period* if

- the *Supervisor* has not done the test or inspection and
- the delay to the test or inspection is not the *Contractor*'s fault.

40.6 The *Project Manager* assesses the cost incurred by the *Employer* in repeating a test or inspection after a Defect is found. The *Contractor* pays the amount assessed.

Testing and inspection before delivery **41**

41.1 The *Contractor* may not bring Plant and Materials to the Working Areas which the Works Information states are to be tested or inspected before delivery until the *Supervisor* has notified the *Contractor* that they have passed the test or inspection.

Searching and notifying Defects **42**

42.1 The *Supervisor* may instruct the *Contractor* to search. He gives his reason for the search with his instruction. Searching may include

- uncovering, dismantling, re-covering and re-erecting work,
- providing facilities, materials and samples for tests and inspections done by the *Supervisor* and
- doing tests and inspections which the Works Information does not require.

42.2 The *Contractor* notifies the *Supervisor* of each Defect which he finds. The *Supervisor* notifies the *Contractor* of each Defect which he finds. No Defects are notified after the *defects date*.

Correcting Defects **43**

43.1 The *Contractor* corrects Defects whether or not the *Supervisor* notifies him of them. The *Contractor* corrects notified Defects before the end of the *defect correction period*. This period begins at Completion for Defects notified before Completion and when the Defect is notified for other Defects.

43.2 The *Supervisor* issues the Defects Certificate at the later of the *defects date* and the end of the last *defect correction period*.

43.3   The *Project Manager* makes arrangements for the *Employer* to give the *Contractor* the access to and use of those parts of the Site which the *Contractor* needs to correct a Defect after the *Employer* has taken over any part of the *works*. If the *Project Manager* has not arranged suitable access and use within the *defect correction period*, he extends the period for the Defect as necessary.

Accepting Defects 44

44.1   The *Contractor* or the *Project Manager* may propose to the other that the Works Information should be changed so that a Defect does not have to be corrected.

44.2   If the *Contractor* and the *Project Manager* are prepared to consider the change, the *Contractor* submits a quotation for reduced Prices or an earlier Completion Date or both to the *Project Manager* for acceptance. If the *Project Manager* accepts the quotation, he gives an instruction to change the Works Information, the Prices and the Completion Date accordingly.

Uncorrected Defects 45

45.1   If the *Contractor* has not corrected a Defect within its *defect correction period*, the *Project Manager* assesses the cost of having the Defect corrected by other people and the *Contractor* pays this amount.

5 Payment

Assessing the amount **50**
due 50.1 The *Project Manager* assesses the amount due at each assessment date. The first assessment date is decided by the *Project Manager* to suit the procedures of the Parties and is not later than the *assessment interval* after the *starting date*. Later assessment dates occur

- at the end of each *assessment interval* until Completion of the whole of the *works*,
- at Completion of the whole of the works,
- when the *Supervisor* issues the Defects Certificate and
- after Completion of the whole of the *works*,
  - when an amount due is corrected and
  - when a payment is made late.

50.2 The amount due is the Price for Work Done to Date plus other amounts to be paid to the *Contractor* less amounts to be paid by or retained from the *Contractor*. Any value added tax or sales tax which the law requires the *Employer* to pay to the *Contractor* is included in the amount due.

50.3 If there is no Accepted Programme and the *Contractor* has not submitted a programme to the *Project Manager* for acceptance which contains the information which this contract requires, half the Price for Work Done to Date is retained from the *Contractor* in assessments of the amount due.

50.4 In assessing the amount due, the *Project Manager* considers any application for payment which the *Contractor* has submitted on or before the assessment date. The *Project Manager* gives the *Contractor* details of how the amount due has been made up.

50.5 The *Project Manager* corrects any wrongly assessed amount due in a later payment certificate.

Payment **51**

51.1 The *Project Manager* certifies a payment within one week of each assessment date. The first payment is the amount due, other payments are the change in the amount due since the last payment certificate. A payment is made by the *Contractor* to the *Employer* if the change reduces the amount due. Other payments are made by the *Employer* to the *Contractor*. Payments are in the *currency of this contract* unless otherwise stated in this contract.

51.2 Each certified payment is made within four weeks of the assessment date or, if a different period is stated in the Contract Data, within the period stated. If a payment is late, interest is paid on the late payment. Interest is assessed from the date by which the late payment should have been made until the date when the late payment is made and is included in the first assessment after the late payment is made.

51.3 If an amount due is corrected in a later certificate either by the *Project Manager* in relation to a compensation event or following a decision of the *Adjudicator* or an arbitrator, interest on the correcting amount is paid. Interest is assessed from the date that the incorrect amount was certified until the date that the correcting amount is certified and is included in the assessment which includes the correcting amount.

51.4 If the *Project Manager* does not issue a certificate which he should issue, interest is paid on the amount which he should have certified. Interest is assessed from the date by which he should have certified the amount until the date when he certifies the amount and is included in the amount then certified.

51.5 Interest is calculated at the *interest rate* and is compounded over the period for which interest is assessed.

Actual Cost **52**

52.1 All the *Contractor*'s costs which are not included in the Actual Cost are deemed to be included in the *fee percentage*. Payments by the *Contractor* included in the Actual Cost are at open market or competitively tendered prices with all discounts, rebates and taxes which can be recovered deducted.

The *bill of quantities* **55**

55.1 **Information in the *bill of quantities* is not Works Information or Site Information.**

6 Compensation events

Compensation events **60**

60.1   The following are compensation events.

(1) The *Project Manager* gives an instruction changing the Works Information except

- a change made in order to accept a Defect or
- a change to the Works Information provided by the *Contractor* for his design which is made at his request or to comply with other Works Information provided by the *Employer*.

(2) The *Employer* does not give possession of a part of the Site by the later of its *possession date* and the date shown on the Accepted Programme.

(3) The *Employer* does not provide something which he is to provide by the date for providing it shown on the Accepted Programme.

(4) The *Project Manager* gives an instruction to stop or not to start any work.

(5) Others do not work within the times shown on the Accepted Programme or do not work within the conditions stated in the Works Information.

(6) The *Project Manager* or the *Supervisor* does not reply to a communication from the *Contractor* within the period required by this contract.

(7) The *Project Manager* gives an instruction for dealing with an object of value or of historical or other interest found within the Site.

(8) The *Project Manager* or the *Supervisor* changes a decision which he has previously communicated to the *Contractor*.

(9) The *Project Manager* withholds an acceptance (other than an acceptance of a quotation for acceleration or for accepting a Defect) for a reason not stated in this contract.

(10) The *Supervisor* instructs the *Contractor* to search and no Defect is found except when the *Contractor* did not notify the *Supervisor* in time for a required test or inspection to be arranged and done before doing work which obstructed the test or inspection.

(11) A test or inspection done by the *Supervisor* causes unnecessary delay.

(12) The *Contractor* encounters physical conditions within the Site, other than weather conditions, which, at the Contract Date, an experienced contractor would have judged to have such a small chance of occurring that it would have been unreasonable for him to have allowed for them.

(13) Weather is recorded within a calendar month and before the Completion Date for the whole of the *works* at the place stated in the Contract Data which one of the *weather measurements*, when compared with the *weather data,* shows has occurred on average less frequently than once in ten years.

(14) An *Employer*'s risk event occurs.

(15) The *Employer* uses part of the *works* before both Completion and the Completion Date.

(16) The *Employer* does not provide materials, facilities and samples for tests as stated in the Works Information.

60.2 In judging the physical conditions, the *Contractor* is assumed to have taken into account

- the Site Information,
- publicly available information referred to in the Site Information,
- information obtainable from a visual inspection of the Site and
- other information which an experienced contractor could reasonably be expected to have or to obtain.

60.3 If there is an inconsistency within the Site Information (including the information referred to in it), the *Contractor* is assumed to have taken into account the physical conditions more favourable to doing the work.

60.4 A difference between the final total quantity of work done and the quantity stated for an item in the *bill of quantities* at the Contract Date is a compensation event if

- **the difference causes the Actual Cost per unit of quantity to change and**
- **the rate in the *bill of quantities* for the item at the Contract Date multiplied by the quantity of work done is more than 0.1 % of the total of the Prices at the Contract Date.**

If the Actual Cost per unit of quantity is reduced, the affected rate is reduced.

60.5 A difference between the final total quantity of work done and the quantity for an item stated in the *bill of quantities* at the Contract Date which delays Completion is a compensation event.

60.6 The *Project Manager* corrects mistakes in the *bill of quantities* which are departures from the *method of measurement* or are due to ambiguities or inconsistencies. Each such correction is a compensation event which may lead to reduced Prices.

Notifying compensation events 61

61.1 For compensation events which arise from the *Project Manager* or the *Supervisor* giving an instruction or changing an earlier decision, the *Project Manager* notifies the *Contractor* of the compensation event at the time of the event. He also instructs the *Contractor* to submit quotations unless the event arises from a fault of the *Contractor* or quotations have already been submitted. The *Contractor* puts the instruction or changed decision into effect.

61.2 The *Project Manager* may instruct the *Contractor* to submit quotations for a proposed instruction or changed decision. The *Contractor* does not put a proposed instruction or changed decision into effect.

61.3 For compensation events not notified by the *Project Manager*, the *Contractor* notifies the *Project Manager* why he considers that the compensation event has happened or is expected to happen. The *Contractor* may not notify a compensation event more than two weeks after he became aware of it.

61.4 For compensation events notified by the *Contractor*, the *Project Manager* decides

- whether the compensation event

  - arises from a fault of the *Contractor*
  - has happened or should be expected to happen
  - could have an effect upon Actual Cost or Completion and

- whether the *Contractor* gave an early warning of the compensation event which an experienced contractor could have given.

61.5 The *Project Manager* notifies the *Contractor* of his decisions within one week of the *Contractor's* notification. If he decides that the compensation event did not arise from a fault of the *Contractor*, that it has happened or should be expected to happen and that it could have an effect upon Actual Cost or Completion, he instructs the *Contractor* to submit quotations at the time he notifies his decisions. If he decides otherwise, the notification of the compensation event is deemed to have been withdrawn.

61.6   If the *Project Manager* decides that the nature of a compensation event is too uncertain for its effect to be forecast reasonably, he states assumptions about its nature on which assessment is to be based when he instructs the *Contractor* to submit quotations. If any of these assumptions is later found to have been wrong, the *Project Manager* notifies a correction as a compensation event.

61.7   A compensation event may not be notified after the *defects date*.

Quotations for **62**
compensation events 62.1   The *Project Manager* may instruct the *Contractor* to submit alternative quotations based upon different ways of dealing with the compensation event which are practicable. The *Contractor* submits the required quotations to the *Project Manager* and may submit quotations for other methods of dealing with the compensation event which he considers practicable.

62.2   Quotations for compensation events comprise proposed changes to the Prices and any delay to the Completion Date assessed by the *Contractor*. He submits details of his assessment with each quotation. If the programme for remaining work is affected by the compensation event, the *Contractor* includes a revised programme showing the effect in his quotation.

62.3   The *Contractor* submits quotations within two weeks of being instructed to do so by the *Project Manager*. The *Project Manager* replies within two weeks of the submission. His reply is

- an instruction to submit a revised quotation,
- an acceptance of a quotation,
- a notification that a proposed instruction or changed decision will not be notified or
- a notification that he will be making his own assessment.

62.4   The *Project Manager* instructs the *Contractor* to submit a revised quotation only after explaining his reasons for doing so to the *Contractor*. The *Contractor* submits the revised quotation within two weeks of being instructed to do so.

Assessing compensation **63**
events 63.1   The changes to the Prices are assessed as the effect of the compensation event upon

- the Actual Cost of the work already done,
- the forecast Actual Cost of the work not yet done and
- the resulting Fee.

63.2   If the effect of a compensation event is to reduce the total Actual Cost, the Prices are not reduced except as stated in this contract. If the effect of a compensation event which is a change to the Works Information or a correction of an assumption stated by the *Project Manager* for assessing an earlier compensation event is to reduce the total Actual Cost, the Prices are reduced.

63.3   A delay to the Completion Date is assessed as the length of time that, due to the compensation event, planned Completion is later than planned Completion as shown on the Accepted Programme.

63.4 If the *Project Manager* has notified the *Contractor* of his decision that the *Contractor* did not give an early warning of a compensation event which an experienced contractor could have given, the compensation event is assessed taking into account any savings of Actual Cost and time which would have occurred if the *Contractor* had given early warning.

63.5 Allowances for cost-increasing and delaying factors which have a significant chance of occurring and which are at the *Contractor*'s risk under this contract are included in forecasts of Actual Cost and Completion.

63.6 Assessments are based upon the assumptions that the *Contractor* reacts competently and promptly to the compensation event, that the additional Actual Cost and time due to the event are reasonably incurred and that the Accepted Programme can be changed.

63.7 A compensation event which is an instruction to change the Works Information in order to resolve an ambiguity or inconsistency is assessed as follows. If Works Information provided by the *Employer* is changed, the effect of the compensation event is assessed as if the Prices and the Completion Date were for the interpretation most favourable to the *Contractor*. If Works Information provided by the *Contractor* is changed, the effect of the compensation event is assessed as if the Prices and the Completion Date were for the interpretation most favourable to the *Employer*.

63.9 Assessments for changed Prices for compensation events are in the form of changes to the *bill of quantities*.

63.10 Assessment of a compensation event for work which includes subcontracted work does not include fees paid or to be paid by the *Contractor* to Subcontractors.

63.11 The *Project Manager* may instruct the *Contractor* to assess the effect of a compensation event upon Actual Cost using the short method set out in the Schedule of Cost Components if the *Contractor* agrees and may make his own assessments using the short method.

The *Project Manager*'s 64
assessments 64.1 The *Project Manager* may assess a compensation event

- if the *Contractor* has not submitted a required quotation and details of his assessment within the time allowed,
- if the *Project Manager* decides that the *Contractor* has not assessed the compensation event correctly in a quotation which is otherwise acceptable,
- if, at the time of notifying the compensation event, the *Contractor* has not submitted a programme which this contract requires him to submit or
- if the reason for the *Project Manager* not accepting the latest programme submitted by the *Contractor* is one of the reasons stated in this contract.

64.2 If there is no Accepted Programme or if the Accepted Programme is out of date, the *Project Manager* assesses the compensation event using his own assessment of the programme for the remaining work.

64.3 The *Project Manager* notifies the *Contractor* of his assessment of a compensation event and gives him details of it within two weeks of the need for his assessment becoming apparent.

Implementing 65
compensation events 65.1 The *Project Manager* implements each compensation event by notifying the *Contractor* of the quotation which he has accepted or of his own assessment. He implements the compensation event when he accepts a quotation or

completes his own assessment or when the compensation event occurs, whichever is latest.

65.2   The assessment of a compensation event is not reviewed in the light of information about its actual effect upon Actual Cost or upon the timing of the work which becomes available only after the *Contractor* has or should have submitted his quotations.

65.4   The *Project Manager* includes the changes to the Prices and the Completion Date from the quotation which he has accepted or from his own assessment in his notification implementing a compensation event.

7 Title

The *Employer*'s title to Equipment, Plant and Materials	**70**	
	70.1	Whatever title the *Contractor* has to Equipment or Plant and Materials which is outside the Working Areas passes to the *Employer* if the *Employer* has made a payment for the Equipment or Plant and Materials or the *Supervisor* has marked it as for this contract.
	70.2	Whatever title the *Contractor* has to Equipment or Plant and Materials passes to the *Employer* if it has been brought within the Working Areas and passes back to the *Contractor* if it is removed from the Working Areas with the permission of the *Project Manager*.
Plant and Materials outside the Working Areas	**71**	
	71.1	The *Supervisor* marks Plant and Materials which are outside the Working Areas if this contract provides for payment for them and the *Contractor* has shown him that any requirements of the Works Information with respect to marking have been satisfied.
Removing Equipment	**72**	
	72.1	The *Contractor* removes Equipment from the Site when it is no longer needed unless the *Project Manager* allows it to be left in the *works*.
Objects and materials within the Site	**73**	
	73.1	The *Contractor* has no title to an object of value or of historical or other interest within the Site. The *Contractor* notifies the *Project Manager* when such an object is found and the *Project Manager* instructs the *Contractor* how to deal with it. The *Contractor* may not move the object without instructions.
	73.2	The *Contractor* has title to materials from excavation and demolition removed from the Site only as stated in the Works Information.

8 Risks and insurance

Employer's and
Contractor's risks

80

80.1 The _Employer_ carries the risks which this contract states are _Employer_'s risks and the _Contractor_ carries the risks which it states are _Contractor_'s risks.

80.2 From the _starting date_ until the Defects Certificate has been issued, the following are _Employer_'s risks.

- The risk of claims, proceedings, compensation and costs payable for personal injury, death or loss of or damage to property (excluding the _works_, Plant and Materials) which are due to

    - use or occupation of the Site by the _works_ or for the purpose of the _works_ which is the unavoidable result of the _works_ or
    - negligence, breach of statutory duty or interference with any legal right by the _Employer_ or by any person employed by or contracted to him except the _Contractor_.

- The risk of damage to the _works_, Plant and Materials to the extent that it is due to a fault of the _Employer_ or in his design.
- The risks of war and of radioactive contamination.
- Other _Employer_'s risks stated in the Contract Data.

80.3 From Risk Transfer until the Defects Certificate has been issued, the risk of loss of or damage to the _works_, Plant and Materials is an _Employer_'s risk except loss or damage due to

- a Defect which existed at Risk Transfer,
- an event occurring before Risk Transfer which was not itself an _Employer_'s risk or
- the activities of the _Contractor_ on the Site after Risk Transfer.

The _Contractor_'s risks

81

81.1 From the _starting date_ until the Defects Certificate has been issued the risks of personal injury, death and loss of or damage to property which are not _Employer_'s risks are _Contractor_'s risks.

Repairs

82

82.1 Before the Defects Certificate has been issued the _Project Manager_ may instruct the _Contractor_ to replace loss of, or repair damage to, the _works_, Plant and Materials due to an _Employer_'s risk event.

82.2 The _Contractor_ promptly replaces loss of and repairs damage to the _works_, Plant and Materials due to a _Contractor_'s risk event.

Liability for personal
injury, death and loss of
or damage to property

83

83.1 Each Party indemnifies the other against claims, compensation and costs for personal injury, death and loss of or damage to property (excluding the _works_, Plant and Materials) due to an event which is at his risk.

83.2 The liability of each Party to indemnify the other is reduced in proportion to the extent that events which were at the other Party's risk contributed to the loss, damage, personal injury or death and taking into account each Party's responsibilities under this contract.

Insurance cover 84

84.1 The *Contractor* provides the following insurances and those stated in the Contract Data unless it is stated in the Contract Data that the *Employer* provides them. The insurances are in the joint names of the Parties and provide cover from the *starting date* until the Defects Certificate has been issued for events which are due to the *Contractor*'s risks. The cover stated is the minimum for any one event.

1. Event: loss of or damage to the *works,* Plant and Materials.
   Cover: 115% of the replacement cost on a value indemnity basis with the maximum deductible stated in the Contract Data and with the amount stated in the Contract Data in respect of the *Contractor*'s faulty design.

2. Event: loss of or damage to Equipment.
   Cover: the replacement cost on a value indemnity basis with the maximum deductible stated in the Contract Data.

3. Event: loss of or damage to property (except the *works*, Plant and Materials and Equipment) in connection with this contract.
   Cover: the amount stated in the Contract Data with cross liability so that the insurance applies to the Parties as separate insured and with the amount stated in the Contract Data in respect of the *Contractor*'s faulty design.

4. Event: personal injury or death.
   Cover: the amounts stated in the Contract Data with cross liability so that the insurance applies to the Parties as separate insured.

84.2 Insurance policies may exclude cover for

- the cost of correcting a Defect,
- consequential loss,
- payment of delay or low performance damages,
- wear and tear, shortages and pilferage and
- risks relating to Equipment which the law requires to be insured.

Insurance policies 85

85.1 The *Contractor* submits policies and certificates for insurance to the *Project Manager* for acceptance before the *starting date* and afterwards as the *Project Manager* instructs. A reason for not accepting the policies and certificates is that they do not comply with this contract.

85.2 Insurance policies include a waiver by the insurers of their subrogation rights against directors and other employees of every insured except where there is fraud.

85.3 The Parties comply with the terms and conditions of the insurance policies.

85.4 Any amount not recovered from an insurer is borne by the *Employer* for his risks or by the *Contractor* for his.

If the *Contractor* does not insure 86

86.1 The *Employer* may insure a risk which this contract requires the *Contractor* to insure if the *Contractor* does not submit a required policy or certificate. The cost of this insurance to the *Employer* is paid by the *Contractor*.

Insurance by the *Employer* 87

87.1 The *Employer* provides the insurances stated in the Contract Data.

87.2   The *Project Manager* submits policies and certificates for insurances provided by the *Employer* to the *Contractor* for acceptance before the *starting date* and afterwards as the *Contractor* instructs. The *Contractor* accepts the policies and certificates if they comply with this contract.

9 Disputes and termination

Disputes about an action taken by the *Project Manager* or the *Supervisor*

90

90.1 If the *Contractor* believes that an action of the *Project Manager* or the *Supervisor* was not in accordance with this contract or was outside the authority given by this contract, he may notify the *Adjudicator* and the *Project Manager* of the disputed action within four weeks of the action. Unless this contract has been terminated, the Parties implement the disputed action.

90.2 Within two weeks of the notification the *Project Manager* provides the *Adjudicator* and the *Contractor* with the information upon which the disputed action was based. Within two weeks of receiving this information the *Contractor* may provide the *Adjudicator* and the *Project Manager* with any other information upon which he believes the *Project Manager* or the *Supervisor* should have based the disputed action.

90.3 The *Adjudicator* decides whether the disputed action was in accordance with this contract and whether it was within the authority given by this contract. If he decides that it was not, he decides what action should have been taken and assesses any additional cost and delay which the dispute itself has caused or will cause to the *Contractor*. The *Adjudicator* makes his assessment in the same way as a compensation event is assessed.

Disputes about an action not taken by the *Project Manager* or the *Supervisor*

91

91.1 If the *Contractor* believes that the *Project Manager* or the *Supervisor* has not taken an action which this contract requires, he may notify the *Project Manager*.

91.2 If the action has not been taken within four weeks of this notification, the *Contractor* may notify the *Adjudicator* and the *Project Manager* within a further four weeks. The *Contractor* may include in this notification information which he believes shows that the *Project Manager* or the *Supervisor* should have taken the action. Within two weeks of the notification to the *Adjudicator*, the *Project Manager* may supply the *Adjudicator* with information which he believes shows that the *Project Manager* or the *Supervisor* should not have taken the action.

91.3 The *Adjudicator* decides whether, in accordance with this contract, the action should or should not have been taken. If he decides that it should have been taken, the action is implemented and he assesses any additional cost and delay which the dispute itself has caused or will cause to the *Contractor*. The *Adjudicator* makes his assessment in the same way as a compensation event is assessed.

The *Adjudicator*

92

92.1 The *Adjudicator* notifies the *Project Manager* and the *Contractor* of his decision, of the reason for his decision and of any assessment within four weeks of receiving the information or within a longer period which has been agreed by the *Project Manager* and the *Contractor*. The *Project Manager* implements the *Adjudicator*'s assessment as if it had resulted from a compensation event.

92.2 If the *Adjudicator* resigns or is unable to act, the Parties choose a new adjudicator jointly. If the Parties have not chosen a new adjudicator jointly within four weeks of the *Adjudicator* resigning or becoming unable to act, any Party may ask the person stated in the Contract Data to choose a new adjudicator and the Parties accept his choice. The new adjudicator is appointed on terms which provide that payments to him are shared equally between the Parties and that he is not liable to the Parties for breach of duty.

92.3 The *Contractor* may notify a dispute arising from a subcontract which is also a dispute under this contract to the *Adjudicator*. The *Adjudicator* then decides the two disputes as one, the Subcontractor becomes a party to the dispute and references to the Parties in clauses of the *conditions of contract* concerned with the dispute are interpreted as including the Subcontractor.

Arbitration 93

93.1 If the *Adjudicator* does not notify his decision within the time provided by this contract or if a Party disagrees with his decision, any Party to the dispute may give to the other parties a notice to refer the dispute to an arbitrator. A notice to refer a dispute to an arbitrator may not be given more than four weeks after the *Adjudicator* has or should have notified his decision.

93.2 A dispute referred to an arbitrator is conducted using the *arbitration procedure*.

Termination 94

94.1 If either Party wishes to terminate, he notifies the *Project Manager* giving details of his reason for terminating. The *Project Manager* issues a termination certificate promptly if the reason complies with this contract.

94.2 The *Contractor* may terminate only for a reason identified in the table. The *Employer* may terminate for any reason. The procedures followed and the amounts due on termination depend upon which Party terminates and upon his reason for terminating, which are identified in the table.

Terminating Party	Reason	Procedure	Amount due
The *Employer*	A reason other than R1 - R18	P1	A1 and A3
	R1 - R12, R16	P1 and P2	A2
	R14, R15, R18	P2	A1 and A4
The *Contractor*	R1 - R7, R13, R17	P3	A1 and A3
	R14, R15, R18	P2	A1 and A4

94.3 The procedures for termination are implemented immediately after the *Project Manager* has issued a termination certificate.

94.4 As soon as possible after termination, the *Project Manager* certifies a final payment to or from the *Contractor* which is the *Project Manager's* assessment of the amount due on termination less the total of previous payments.

94.5 After a termination certificate has been issued, the *Contractor* does no further work necessary to complete the *works*.

Reasons for termination 95

95.1 Either Party may terminate if the other Party has done one of the following or its equivalent

- become bankrupt or insolvent (R1),
- had a bankruptcy order made against him (R2),
- presented his petition in bankruptcy (R3),
- made an arrangement with or an assignment in favour of his creditors (R4),

- agreed to perform this contract under a committee of inspection of his creditors (R5),
- gone into liquidation other than voluntarily in order to amalgamate or reconstruct (R6) or
- assigned this contract (R7).

95.2   The *Employer* may terminate if the *Project Manager* has notified that the *Contractor* has defaulted in one of the following ways and not put the default right within four weeks of the notification.

- Substantially failed to comply with his obligations (R8).
- Not provided a bond or guarantee which this contract requires (R9).
- Appointed a Subcontractor for substantial work before the *Project Manager* has accepted the Subcontractor (R10).

95.3   The *Employer* may terminate if the *Project Manager* has notified that the *Contractor* has defaulted in one of the following ways and not stopped defaulting within four weeks of the notification.

- Substantially hindered the *Employer* or Others (R11).
- Substantially broken a health or safety regulation (R12).

95.4   The *Contractor* may terminate if the *Employer* has not paid an amount certified by the *Project Manager* within thirteen weeks of the date of the certificate (R13).

95.5   Either Party may terminate if

- war or radioactive contamination has substantially affected the *Contractor*'s work for 26 weeks (R14) or
- the Parties have been released under the law from further performance of the whole of this contract (R15).

95.6   If the *Project Manager* has instructed the *Contractor* to stop or not to start any substantial work or all work and an instruction allowing the work to restart or start has not been given within thirteen weeks

- the *Employer* may terminate if the instruction was due to a default by the *Contractor* (R16)
- the *Contractor* may terminate if the instruction was due to a default by the *Employer* (R17) and
- either Party may terminate if the instruction was due to any other reason (R18).

Procedures on   96
termination   96.1   The *Employer* may complete the *works* himself or employ other people to do so and may use any Plant and Materials to which he has title.

96.2   The procedure on termination also includes one or more of the following as set out in the table.

P1	The *Employer* may instruct the *Contractor* to leave the Site, remove any Equipment, Plant and Materials from the Site and assign the benefit of any subcontract or other contract related to performance of this contract to the *Employer*.
P2	The *Employer* may use any Equipment to which he has title.
P3	The *Contractor* leaves the Working Areas and removes the Equipment.

Payment on termination **97**

97.1   The amount due on termination includes

- an amount due assessed as for normal payments,
- the Actual Cost for Plant and Materials

    - within the Working Areas or
    - to which the *Employer* has title and of which the *Contractor* has to accept delivery,

- other Actual Cost reasonably incurred in expectation of completing the *works*,
- any amounts retained by the *Employer* and
- a deduction of any unrepaid balance of an advanced payment.

97.2   The amount due on termination also includes one or more of the following as set out in the table.

A1	The forecast Actual Cost of removing the Equipment.
A2	A deduction of the forecast of the additional cost to the *Employer* of completing the *works*.
A3	The *fee percentage* applied to • for Options A, B, C and D, any excess of the total of the Prices at the Contract Date over the Price for Work Done to Date or • for Options E and F, any excess of the first forecast of the Actual Cost for the *works* over the Price for Work Done to Date less the Fee.
A4	Half of A3.

SECONDARY OPTION CLAUSES

Option G : Performance bond

Performance bond **G1**

G1.1   A bank or insurer which the *Project Manager* has accepted gives the *Employer* a performance bond in the form set out in the Works Information and for the amount stated in the Contract Data. If the bond was not given by the Contract Date, it is given to the *Employer* within four weeks of the Contract Date. A reason for not accepting the bank or insurer is that its commercial position is not strong enough to carry the bond.

Option H : Parent company guarantee

Parent company **H1**
guarantee H1.1   If a parent company owns the *Contractor*, the parent company gives a guarantee of the *Contractor*'s performance to the *Employer* in the form set out in the Works Information. If it was not given by the Contract Date, it is given within four weeks of the Contract Date.

Option J : Advanced payment to the *Contractor*

Advanced payment **J1**

J1.1   The *Employer* makes an advanced payment to the *Contractor* of the amount stated in the Contract Data.

J1.2   The advanced payment is made either within four weeks of the Contract Date or, if an advanced payment bond is required but has not been given before the Contract Date, within four weeks of a bank or insurer which the *Project Manager* has accepted giving an advanced payment bond to the *Employer*. A reason for not accepting the proposed bank or insurer is that its commercial position is not strong enough to carry the bond. The bond is for the amount of the advanced payment and in the form set out in the Works Information. Delayed payment of the advanced payment is a compensation event.

J1.3   The advanced payment is repaid to the *Employer* by the *Contractor* in instalments of the amount stated in the Contract Data. An additional instalment is included in each amount due assessed after the period stated in the Contract Data has passed until the advanced payment has been repaid.

Option K : Multiple currencies

Multiple currencies **K1**

K1.1   The *Contractor* is paid in currencies other than the *currency of this contract* for the work listed in the Contract Data. The *exchange rates* are used to convert from the *currency of this contract* to other currencies.

K1.2   Payments to the *Contractor* in currencies other than the *currency of this contract* do not exceed the maximum amounts stated in the Contract Data. Any excess is paid in the *currency of this contract*.

Option L : Sectional Completion

Sectional Completion **L1**

L1.1   Each reference in the *conditions of contract* to the *works*, to Completion and to the Completion Date applies to the *works* and any *section* of the *works* unless it is stated to apply to the whole of the *works*.

Option M : Limitation of the *Contractor*'s liability for his design to reasonable skill and care

The *Contractor*'s design **M1**

M1.1   The *Contractor* is not liable for Defects in the *works* due to his design so far as he proves that he used reasonable skill and care to ensure that it complied with the Works Information.

Option N : Price adjustment for inflation

Defined terms **N1**

N1.1   (a) The Base Date Index (B) is the latest available index before the *base date*.

(b) The Latest Index (L) is the latest available index before the date of assessment of an amount due.

(c) The Price Adjustment Factor is the total of the products of each of the proportions stated in the Contract Data multiplied by $(L - B)/B$ for the index linked to it.

Price Adjustment Factors N2

N2.1   If an index is changed after it has been used in calculating a Price Adjustment Factor, the calculation is repeated and a correction included in the next assessment of the amount due.

N2.2   The Price Adjustment Factor calculated at the Completion Date for the whole of the *works* is used for calculating price adjustment after this date.

Compensation events N3

N3.1   The Actual Cost for compensation events is assessed using the

* Actual Costs current at the time of assessing the compensation event adjusted to *base date* by dividing by one plus the Price Adjustment Factor for the last assessment of the amount due and
* Actual Costs at *base date* levels for amounts calculated from rates stated in the Contract Data for employees and Equipment.

Price adjustment N4

N4.1   Each amount due includes an amount for price adjustment which is the sum of

* the change in the Price for Work Done to Date since the last assessment of the amount due multiplied by the Price Adjustment Factor for the date of the current assessment,
* the amount for price adjustment included in the previous amount due and
* correcting amounts, not included elsewhere, which arise from changes to indices used for assessing previous amounts for price adjustment.

Option P : Retention

Retention P1

P1.1   Nothing is retained until the Price for Work Done to Date has reached the *retention free amount*. An amount is then retained from the *Contractor* in each amount due until the earlier of Completion of the whole of the *works* and the *Employer* taking over the whole of the *works*. This amount is the *retention percentage* applied to the excess of the Price for Work Done to Date above the *retention free amount*.

P1.2   The amount retained is halved in the amount due assessed at Completion of the whole of the *works* or in the first assessment after the *Employer* has taken over the whole of the *works* if this is before Completion of the whole of the *works*. The amount retained remains at this amount until the Defects Certificate is issued. Nothing is retained in the assessment made when the Defects Certificate is issued or in later assessments of the amount due.

Option Q: Bonus for early Completion

Bonus for early Completion

Q1

Q1.1   The *Contractor* is paid a bonus calculated at the rate stated in the Contract Data for each day from the earlier of Completion and the date on which the *Employer* begins to use the *works* until the Completion Date.

Option R : Delay damages

Delay damages

R1

R1.1   The *Contractor* pays delay damages at the rate stated in the Contract Data for each day from the Completion Date until the earlier of Completion and the date on which the *Employer* begins to use the *works*.

R1.2   If the Completion Date is delayed after delay damages have been paid, the *Employer* repays the overpayment of damages with interest. Interest is assessed from the date of payment to the date of repayment and the date of repayment is an assessment date.

Option S : Low performance damages

Low performance damages

S1

S1.1   If a Defect included in the Defects Certificate shows low performance with respect to a performance level stated in the Contract Data, the *Contractor* pays the amount of low performance damages stated in the Contract Data.

Option T : Changes in the law

Changes in the law

T1

T1.1   A change in the law of the country in which the Site is located is a compensation event if it occurs after the Contract Date. The *Project Manager* may notify the *Contractor* of a compensation event for a change in the law and instruct him to submit quotations. If the effect of a compensation event which is a change in the law is to reduce the total Actual Cost, the Prices are reduced.

Option U : Special conditions of contract

Special conditions of   **U1**
contract   U1.1   The special conditions of contract stated in the Contract Data are part of this contract.

SCHEDULE OF COST COMPONENTS

Amounts are included only in one cost component.

People 1 The following components of the cost of

- people who are directly employed by the *Contractor* and either working within the Working Areas or whose normal place of working is within the Working Areas and
- people who are not directly employed by the *Contractor* but are paid by the *Contractor* according to the time worked whilst they are within the Working Areas.

11 Wages and salaries.

12 Payments to people for

(a) bonuses and incentives
(b) overtime
(c) working in special circumstances
(d) special allowances
(e) absence due to sickness and holidays
(f) severance related to work on this contract.

13 Payments made in relation to people for

(a) travelling to and from the Working Areas
(b) subsistence and lodging
(c) relocation
(d) medical examinations
(e) passports and visas
(f) travel insurance
(g) items (a) to (f) for a spouse or dependents
(h) protective clothing
(j) meeting the requirements of the law
(k) superannuation and life assurance
(l) death benefit
(m) occupational accident benefits
(n) medical aid.

Equipment 2 The following components of the cost of Equipment which is used within the Working Areas (excluding Equipment cost covered by the percentage for Working Areas overheads).

21 Payments for hire of Equipment not owned by the *Contractor*, by the *Contractor*'s parent company or by another part of a group with the same parent company.

22 An amount for depreciation and maintenance of Equipment which is

(a) owned by the *Contractor*,
(b) purchased by the *Contractor* under a hire purchase or lease agreement or
(c) hired by the *Contractor* from the *Contractor*'s parent company or another part of a group with the same parent company.

The depreciation and maintenance charge is the actual purchase price of the item of Equipment (or first cost if the *Contractor* assembled the item) divided by its average working life remaining at the time of purchase (expressed in weeks).

The amount for depreciation and maintenance is calculated by multiplying the depreciation and maintenance charge by the time required (expressed in weeks) and then adding the percentage for Equipment depreciation and maintenance stated in the Contract Data.

23   The purchase price of Equipment which is consumed.

24   Except when covered by hire rates, payments for

(a) transporting Equipment to and from the Working Areas and
(b) erecting and dismantling Equipment.

Plant and Materials   **3**   The following components of the cost of Plant and Materials.

31   Payments for

(a) purchasing Plant and Materials
(b) delivery to and removal from the Working Areas
(c) providing and removing packaging
(d) samples and tests.

32   Cost is credited with payments received for disposal of Plant and Materials.

Charges   **4**   The following components of the cost of charges paid by the *Contractor.*

41   Payments to utilities for provision and use in the Working Areas of

(a) water
(b) gas
(c) electricity
(d) other services.

42   Payments to public authorities, utilities and other properly constituted authorities of charges which they are authorised to make in respect of the *works*.

43   Payments for

(a) financing charges (excluding charges compensated for by interest
paid in accordance with this contract)
(b) buying or leasing land
(c) compensation for loss of crops or buildings
(d) royalties
(e) inspection certificates
(f) rent of premises in the Working Areas
(g) charges for access to the Working Areas
(h) facilities for visits to the Working Areas by Others
(i) specialist services.

44   A charge for overhead costs incurred within the Working Areas calculated by applying the percentage for Working Areas overheads stated in the Contract Data to the Actual Cost of people. The charge is deemed to include provision and use of accommodation, equipment, supplies and services for

(a) offices and drawing offices
(b) laboratories
(c) workshops
(d) stores and compounds
(e) labour camps
(f) cabins
(g) catering
(h) medical facilities and first aid

(j) recreation
(k) sanitation
(l) security
(m) copying
(n) telephone, telex, fax, radio and CCTV
(o) surveying and setting out
(p) computing
(q) tools.

Manufacture and fabrication 5   The following components of the cost of manufacture and fabrication done by the *Contractor* or a Subcontractor outside the Working Areas.

51   The total of the hours worked by employees multiplied by the hourly rates stated in the Contract Data for the categories of employees listed.

52   An amount for overheads calculated by multiplying this total by the percentage for manufacturing and fabrication overheads stated in the Contract Data.

Design 6   The following components of the cost of design of the *works* and Equipment done outside the Working Areas.

61   The total of the hours worked by employees multiplied by the hourly rates stated in the Contract Data for the categories of employees listed.

62   An amount for overheads calculated by multiplying this total by the percentage for design overheads stated in the Contract Data.

63   The cost of travel to and from the Working Areas for the categories of people listed in the Contract Data.

Insurance 7   Insurer's payments of claims are deducted from cost.

SCHEDULE OF COST COMPONENTS USING THE SHORT METHOD

Amounts are included only in one cost component.

People   **1**   The following components of the cost of

- people who are directly employed by the *Contractor* and either working within the Working Areas or whose normal place of working is within the Working Areas and
- people who are not directly employed by the *Contractor* but are paid by the *Contractor* according to the time worked while they are within the Working Areas.

     11   Wages and salaries.

     12   Payments to people for

         (a) bonuses and incentives
         (b) overtime
         (c) working in special circumstances
         (d) special allowances.

     13   Payments made in relation to people for

         (a) travelling to and from the Working Areas
         (b) subsistence and lodging.

     14   A charge for overhead costs for people, payments to Others and Working Area overheads calculated by applying the percentage for people overheads stated in the Contract Data to the total of items 11, 12 and 13.

Equipment   **2**   The following components of the cost of Equipment used within the Working Areas (excluding Equipment cost covered by the percentage for people overheads)

     21   Amounts for Equipment which is in the published list stated in the Contract Data. These amounts are calculated by applying the percentage adjustment for listed equipment stated in the Contract Data to the rates in the published list and by multiplying the resulting rate by the time for which the Equipment is working.

     22   Amounts for Equipment listed in the Contract Data which is not in the published list stated in the Contract Data. These amounts are the rates stated in the Contract Data multiplied by the time for which the Equipment is working.

Plant and Materials   **3**   The following are components of the cost of Plant and Materials.

     31   Payments for

         (a) purchasing Plant and Materials
         (b) delivery to and removal from the Working Areas
         (c) providing and removing packaging
         (d) samples and tests.

     32   Cost is credited with payments received for disposal of Plant and Materials.

Manufacture and fabrication   **5**   The following components of the cost of manufacture and fabrication done by the *Contractor* or a Subcontractor outside the Working Areas.

51   The total of the hours worked by employees multiplied by the hourly rates stated in the Contract Data for the categories of employees listed.

52   An amount for overheads calculated by multiplying this total by the percentage for manufacturing and fabrication overheads stated in the Contract Data.

Design 6   The following components of the cost of design of the *works* and Equipment done outside the Working Areas.

61   The total of the hours worked by employees multiplied by the hourly rates stated in the Contract Data for the categories of employees listed.

62   An amount for overheads calculated by multiplying this total by the percentage for design overheads stated in the Contract Data.

63   The cost of travel to and from the Working Areas for the categories of people listed in the Contract Data.

Insurance 7   Insurer's payments of claims are deducted from cost.

CONTRACT DATA

Part one – Data provided by the Employer

1. General The *conditions of contract* are the core clauses and the clauses for secondary options

. . . . of the New Engineering Contract.

- The *works* are

. . . .

- The *Employer* is

Name . . . .

Address . . . .

. . . .

- The *Project Manager* is

Name . . . .

Address . . . .

. . . .

- The *Supervisor* is

Name . . . .

Address . . . .

. . . .

- The *Adjudicator* is

Name . . . .

Address . . . .

. . . .

- The Works Information and the Site Information are in

. . . .

. . . .

. . . .

. . . .

. . . .

. . . .

. . . .

- The *boundaries of the site* are . . . .

- The *language of this contract* is . . . .

- This contract is governed by the law of ....

- The *period for reply* to a communication is ....weeks

2. The *Contractor*'s main responsibilities
- After Completion of the whole of the *works*, the limit of the *Contractor*'s liability for his design is ....

3. Time
- The *starting date* is ....

- The *possession dates* are

Part of the Site	Date
1	....
2	....
3	....

- The *Contractor* submits revised programmes at intervals no longer than

  .... weeks

4. Testing and Defects
- The *defects date* is .... weeks after Completion of the whole of the *works*

- The *defect correction period* is .... weeks

5. Payment
- The *currency of this contract* is the ....

- The *assessment interval* is .... weeks [not more than five]

- The *interest rate* is .... % per ....(time period) above/below the

  .... rate of the ....

6. Compensation events
- The *weather measurements* are

  - rainfall (mm)..................

  - days with rainfall greater than 5 mm (nr) ....

  - days with minimum air temperature less than 0 degrees Celsius (nr) .....

  - days with snow lying at .... hours GMT (nr) ....

  and these measurements

  ....

  ....

  ....

- The *weather data* are the records of past *weather measurements* which were

  recorded at ....

  and which are available from ....

  ....

  Where no recorded data are available, assumed values for the ten year return monthly *weather data* are stated here.

- The place where weather is to be recorded is ....

8. Risks and insurance
- The maximum deductible for insurance of the *works* and of Plant and Materials is
  . . . .
- The minimum cover for insurance of the *works* and of Plant and Materials in respect of the *Contractor*'s faulty design is . . . .
- The maximum deductible for insurance of Equipment is. . . . .
- The cover for the insurance of other property is . . . .
- The minimum cover for insurance of other property in respect of the *Contractor*'s faulty design is . . . .
- The minimum cover for personal injury or death insurance
  - for the *Contractor*'s employees is . . . .
  - and for other people is . . . .

9. Disputes and termination
- The person who will choose an adjudicator if the Parties cannot agree a choice is
  . . . .
- The *arbitration procedure* is . . . .
  . . . .

- **The *method of measurement* is** . . . .
  amended as follows . . . .
  . . . .
  . . . .

Optional statements

If the Completion Date has been decided by the *Employer*
- The *completion date* for the whole of the *works* is . . . .

If the *Employer* is not willing to take over the *works* before the Completion Date
- The *Employer* is not willing to take over the *works* before the Completion Date

If the *Contractor* is not to submit the Accepted Programme with his tender
- The *Contractor* is to submit the Accepted Programme within . . . . weeks of the Contract Date

If the period for payment is not four weeks
- The period within which payments are made is . . . . .weeks

If there are additional compensation events

- These are compensation events

    1 . . . .

    2 . . . .

    3 . . . .

If there are additional *Employer*'s risks

- These are *Employer*'s risks

    1 . . . .

    2 . . . .

    3 . . . .

If the *Contractor* is to provide additional insurances

- The *Contractor* provides these insurances

Event	Cover
1	. . . .
2	. . . .
3	. . . .

If the *Employer* is to provide insurances

- The *Employer* provides these insurances

Event	Cover
1	. . . .
2	. . . .
3	. . . .

If Option G is used

- The amount of the performance bond is . . . .

If Option J is used

- The amount of the advanced payment is . . . .

- The instalments to be repaid by the *Contractor* in assessments starting not less

    than . . . . weeks after the Contract Date are . . . .

    (either an amount or a percentage of the payment otherwise due)

- An advanced payment bond is/is not required

If Option K is used

- Payment for the items or activities listed below will be made in the currencies stated

Items or activities	Currency	Maximum payment
. . . .	. . . .	. . . .
. . . .	. . . .	. . . .
. . . .	. . . .	. . . .

- The *exchange rates* are those published in . . . .

    on . . . . (date)

If Option L is used

- The *completion date* for each *section* of the *works* is

Section	Description	*Completion date*
1	. . . .	. . . .
2	. . . .	. . . .
3	. . . .	. . . .
4	. . . .	. . . .
5	. . . .	. . . .

If Options L and Q are used together

- The bonuses for the *section*s of the *works* are

Section	Description	Amount per day
1	. . . .	. . . .
2	. . . .	. . . .
3	. . . .	. . . .
4	. . . .	. . . .
5	. . . .	. . . .

If Options L and R are used together

- Delay damages for the *section*s of the *works* are

Section	Description	Amount per day
1	. . . .	. . . .
2	. . . .	. . . .
3	. . . .	. . . .
4	. . . .	. . . .
5	. . . .	. . . .

If Option N is used

- The proportions used to calculate the Price Adjustment Factor are

  0 · . . . .   linked to the index for . . . .

  0 · . . . .

  0 · . . . .

  0 · . . . .

  0 · . . . .

  0 · . . . .

  0 · . . . .   non-adjustable
  _____
  1 · 00

- The *base date* for indices is. . . . .

- The indices are those prepared by . . . .

If Option P is used

- The *retention free amount* is . . . .

- The *retention percentage* is . . . . %

If Option Q is used

- The bonus for the whole of the *works* is . . . . per day

If Option R is used (whether or not Option L is also used)

- Delay damages for the whole of the *works* are . . . . per day

If Option S is used

- The amounts for low performance damages are

Amount	Performance level
. . . .	for . . . .
. . . .	for . . . .
. . . .	for . . . .
. . . .	for . . . .

If Option U is used

- These special conditions apply

  . . . .

  . . . .

Part two - Data provided by the *Contractor*

Statements given in all contracts

- The *Contractor* is

  Name . . . .

  Address . . . .

  . . . .

- The *fee percentage* is . . . .

- The *working areas* are the Site and . . . .

- The key people are

  (1) Name . . . .

  Job . . . .

  Responsibilities . . . .

  . . . .

  Qualifications . . . .

  Experience . . . .

  . . . .

  (2) Name . . . .

  Job . . . .

  Responsibilities . . . .

  . . . .

  Qualifications . . . .

  Experience . . . .

  . . . .

- **The *bill of quantities* is** . . . .

- **The tendered total of the Prices is** . . . .

Optional statements   If the Contractor is to provide Works Information for his design

- The Works Information for the *Contractor*'s design is in

  . . . .

  . . . .

  . . . .

  . . . .

  . . . .

If the *Contractor* is to provide a programme

- The programme numbered . . . . is the Accepted Programme

If the *Contractor* is to provide the *completion date* for the whole of the *works*

- The *completion date* for the whole of the *works* is . . . .

Cost component data

- The percentage for Equipment depreciation and maintenance is. . . . . % (not used for the short method)

- The percentage for Working Areas overheads is . . . . % (not used for the short method)

- The hourly rates for Actual Cost of manufacture and fabrication outside the Working Areas are

- Category of employee   Hourly rate

  . . . .

  . . . .

  . . . .

  . . . .

The percentage for manufacture and fabrication overheads is . . . . %

- The hourly rates for Actual Cost of design outside the Working Areas are

- Category of employee   Hourly rate

  . . . .

  . . . .

  . . . .

  . . . .

- The percentage for design overheads is . . . . %

- The categories of employees whose travelling expenses to and from the Working Areas are included in Actual Cost are

  . . . .

  . . . .

  . . . .

  . . . .

Cost component data for the short method

- The percentage for people overheads is . . . . %

- The published list of Equipment is the last edition of the list published by

  . . . .

- The percentage for adjustment for listed Equipment is . . . . %

- The list of Equipment and rates is

Description	Size or capacity	Rate
. . . .	. . . .	. . . .
. . . .	. . . .	. . . .
. . . .	. . . .	. . . .
. . . .	. . . .	. . . .
. . . .	. . . .	. . . .
. . . .	. . . .	. . . .